Did you enjoy this issue of BioCoder?

Sign up and we'll deliver future issues and news about the community for FREE.

http://oreilly.com/go/biocoder-news

BioCoder

WINTER 2015

Beijing · Cambridge · Farnham · Köln · Sebastopol · Tokyo

BioCoder #6

Copyright © 2015 O'Reilly Media, Inc. All rights reserved.

Printed in the United States of America.

Published by O'Reilly Media, Inc., 1005 Gravenstein Highway North, Sebastopol, CA 95472.

O'Reilly books may be purchased for educational, business, or sales promotional use. Online editions are also available for most titles (*http://safaribooksonline.com*). For more information, contact our corporate/institutional sales department: 800-998-9938 or *corporate@oreilly.com*.

Editors: Mike Loukides and Nina DiPrimio	**Interior Designer:** David Futato
Production Editor: Nicole Shelby	**Cover Designer:** Edie Freeman and Ellie
Copyeditor: Amanda Kersey	Volckhausen
Proofreader: Rachel Head	**Illustrator:** Rebecca Demarest

January 2015:   First Edition

Revision History for the First Edition

2015-01-12   :   First Release

Contents

A Programming Language for Biology

Mike Loukides

In a couple of recent (*http://bit.ly/bio6-anticommons*) posts (*http://bit.ly/bio6-os-bio*), I've written about the need for a high-level programming language for biology. Now we have one. Antha (*http://antha-lang.org/*) is a high-level, open source language for specifying biological workflows (i.e., describing experiments). It's available on GitHub (*https://github.com/antha-lang*).

A programming language for scientific experiments is important for many reasons. Most simply, a scientist in training spends many, many hours of time learning how to do lab work. That sounds impressive, but it really means moving very small amounts of liquid from one place to another. Thousands of times a day, thousands of days in preparation for a career. It's boring, dull, and necessary work, and something that can be automated. Biologists should spend most of their time thinking about biology, designing experiments, and analyzing results—not handling liquids.

More importantly, we've all read reports about experimental results that aren't reproducible (*http://bit.ly/bio6-nature*). That doesn't necessarily mean that the results are invalid, but rather that they're susceptible to small changes in the way the experiment is carried out: a particular person's pipetting technique, the particular lab equipment used, or parts of the experiment that nobody would bother to record in a scientific paper. Describing the process completely, performing the experiments on robotic lab equipment, and automating the data collection not only eliminates many variables but also makes it possible to collect more data than was previously possible: data about every aspect of the experiment. With more data, you might even be able to determine why some trials work and others don't.

We've heard that the use of robotics in research has been limited because it's difficult to program the current generation of robots; programming is very low-level ("move to the right 3.9 cm, move down .45 cm, push this button on the pipette..."). A high-level programming language does more than make the process simpler; it makes it easier to scale from small or moderate-sized experiments to much greater size. 384 samples? Why not 4096? Once you have the programming taken care of and have automated both the experimental process and the data collection, you've eliminated many sources of error and can do much bigger experiments. That's precisely the problem that programming in a high-level language solves. You describe the experiment, and iteration comes for free, at whatever scale you want. Lab robots never get bored or tired.

Antha isn't yet ready for use: the compiler is available, along with some development tools, but drivers for common lab equipment aren't available yet (though several manufacturers are already working on them). However, it's time for biologists to start thinking about it and how to integrate it into their labs. What tools do you want? What devices do you need to automate? What would an environment for designing experiments look like?

It's fascinating to watch the future unfold.

Cataloging Strains: Isolation and Identification of Invasive Fungi on *Citrus limon*

Louis Huang and Alan Rockefeller

This project was sponsored by Counter Culture Labs and conducted in the San Francisco Bay Area. For questions and information on the project, please contact Louis Huang at *louis.huang@aegia.nu*.

Abstract

An unknown mold on a *Citrus limon* tree was isolated and identified using standard microbiology protocols: streaking, microscopy, and DNA barcoding. This was done to identify invasive mold that was hindering the growth of *C. limon* fruit and to catalog its relationship with other similar fungal species. The method details an efficient and cost-effective way to isolate and identify a wild-type organism.

Introduction

The isolation and identification of a wild-type organism infecting *C. limon* was conducted. Due to its characteristics of phenotype and growth behavior—an earthy-colored fuzzy organism growing on expiring fruit—the targeted organism was presumed to be a mold. Using a combination of streaking, microscopy, and DNA barcoding, the strain of invasive mold was identified. The barcode sequence was matched with other genetically similar strains and cataloged in a phylogenetic tree.

The *streaking method* isolates a pure strain microorganism from a diverse population. *Microscopy*, as applied in this experiment, is the use of microscopes to classify the genus of the organism, also known as taxonomy. The method of *DNA barcoding*, also a taxonomic indicator, determines an unidentified organism's species by reading a short region of genetic sequence unique to that species. The *phylogenetic tree*, a certain type of node graph, is used to study the relationships of evolutionary lineages.

The strain found in this study matched several other sequences found in NCBI's database (*http://bit.ly/bio6-ncbi*) of *Penicillium brevicompactum*. The experiment illustrates the efficient use of microbiology techniques that may be used as a standard for isolating and identifying future unknown microorganisms, specifically of the fungus kingdom.

Materials and Methods

ISOLATION

Swab samples were taken of two moldy fruits from the *C. limon* tree. One sample isolated from each of the two fruits and one mixed sample were inoculated on three different plates using Czapek Dox (Cz) agar. After an initial microscopy, each plate was inoculated by streaking method on Czapek Yeast Autolysate (CYA) agar.

Czapek Dox agar, a commonly used media for the cultivation of fungi, was used as the base medium of growth for the targeted unknown organism of the fungi kingdom. Once the mold was initially determined to be of *Penicillium* genus by microscopy, CYA agar was used due to its standard as a monograph for *Penicillium*.[1]

Plates were stored in the dark at room temperature and were subcultured (a technique to isolate a pure strain by streaking on a new plate) every two to three days. Initial samples were inoculated three times. All further inoculations were streaked in the center or three-point inoculated.

Table 2-1. Czapek Dox (Cz) agar composition[2]

$NaNO_3$	0.300%
Sucrose	3.000%
$K_2HPO_4 \cdot 3H_2O$	0.130%

MgSO$_4$·7H$_2$O	0.050%
KCl	0.050%
FeSO$_4$·7H$_2$O	0.001%
CuSO$_4$·5H$_2$O	0.001%
ZnSO$_4$·7H$_2$O	0.001%
Agar	1.500%

Table 2-2. Czapek Yeast Autolysate (CYA) agar composition[3]

NaNO$_3$	0.300%
Yeast extract (Difco)	0.500%
Sucrose	3.000%
K$_2$HPO$_4$·3H$_2$O	0.130%
MgSO$_4$·7H$_2$O	0.050%
KCl	0.050%
FeSO$_4$·7H$_2$O	0.001%
CuSO$_4$·5H$_2$O	0.001%
ZnSO$_4$·7H$_2$O	0.001%
Agar	1.500%

MICROSCOPY

Thin cross sections of isolated sample from initial inoculated and final subcultured plates were viewed under an optical light microscope for identification and confirmation up to the genus level of the unknown organism.

In the initial microscopy, thin strips of fungi inoculated from *C. limon* were viewed microscopically for mold confirmation and identification to genus. Microscopy on the final subcultured plates was done to confirm consistency matching the initial microscopy.

The samples were inspected with methylene blue dye to enhance features of the target from its background. Magnifications of 100x, 400x, and 1000x were sampled on an AmScope M150C-MS 40x–1000x microscope.

DNA BARCODE

The DNA extraction was done by placing 10 mg of unknown mold into 0.5 M NaOH and heated to 80 degrees Celsius for 10 minutes. Afterward, 10 µL of the extract was put into 495 µL of pH 8.0 Tris buffer. One µL of Tris-buffered DNA extract was used as template in a 25 µL PCR reaction that included 22 µL of Gene and Cell master mix, 1 µL of 10 nM ITS1 primer, and 1 µL of 10 nM ITS4 primer. The PCR was programmed for 2 minutes at 95 degrees Celsius, followed by 35 cycles of 95 degrees for 30 seconds, followed by 1 minute of annealing at 55 degrees and 1 minute of extension at 72 degrees. After 35 cycles, the samples were kept at 72 degrees for 5 minutes, then cooled down to 4 degrees. Sequencing was done by Genewiz in Berkeley. The species was identified using NCBI's BLAST.

This study selected the nuclear ribosomal internal transcribed spacer (ITS) gene region for the analysis because of the region's common use in sequence identification of fungi, allowing comparison of our identified sequence with various sequences in GenBank. The primers selected were the forward primer ITS1 (TCCGTAGGTGAACCTGCGG) and reverse primer ITS4 (TCCTCCGCTTATTGATATGC). While the full genome of *Penicillium* contains approximately five million base pairs, the primers search the DNA for portions that are complementary to its primer sequences. When it finds a match, it cuts and terminates the DNA molecule. The target strand is amplified, while the rest of the DNA is not.

PHYLOGENETICS

Relating properly identified sequences in GenBank's database to the targeted organism's barcode sequence by using NCBI's BLAST search, a phylogenetic tree was created. The phylogenetic tree was rendered by the MEGA6 software, which aligned the sequences using the MUSCLE algorithm. The Find Best DNA/ Protein Models feature was used to determine the most accurate mathematical model. A maximum-likelihood tree was created using the the general time reversible model with gamma-distributed rates among sites.

Results

ISOLATION

After 48 hours of growth of the initial sample inoculation, various distinct colors of mold were found to be growing on the plates—primarily a gray-green mold and an olive-colored mold. After the second and third subcultures, the gray-green-colored mold was isolated for this study.

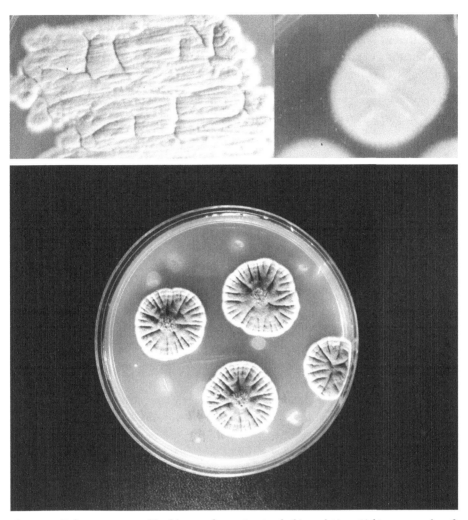

Figure 2-1. Left: gray-green mold 96 hours after center-streaked inoculation. Right: reverse color of target mold. Below: gray-green mold six days after three-point inoculation.

MICROSCOPY

The microscopic features observed were a good match for the genus *Penicillium*. Branched conidiophores, metulae, sterigmata, and conidia were observed. The sterigmata and conidiospores of the unknown mold have the paintbrush-like appearance seen in *Penicillium* species.[4] The conidia (asexual spores) are round and unicellular.

The anatomies of various fungi species were compared to the targeted organism. Different shapes of the conidia grow from the sterigmata, which branches out from the metulae in different genera of fungi. The entire conidia bearing branch and/or hypha is the conidiospore.

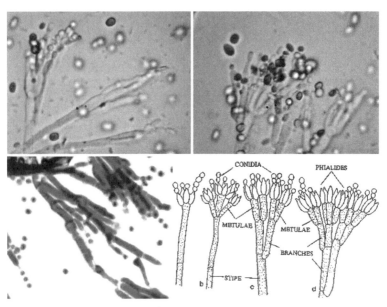

Figure 2-2. Above: sample shows distinctive, brush-like shape of the conidiophores. Magnification of 400x, lightly stained with methylene blue. Below: references of Penicillium from Morgan, 1999, and Samson et al., 1984.

DNA BARCODE

A common locus for DNA barcoding fungi is the internal transcribed spacer (ITS) region,[5] where different locations are used for different kingdoms. The resulting sequence of DNA from the primer cut ITS region was:

```
TCCACCTCCCACCCGTGTTTATTTTACCTTGTTGCTTCGGCGAGCCTGCCTTTTGGCTGCCGGGG-
GACGTCTGTCCCCGGGTCCGCGCTCGCCGAAGACACCTTAGAACTCTGTCTGAAGATTGTAGTCTGAGATTAAATA-
TAAATTATTTAAAACTTTCAACAACGGATCTCTTGGTTCCGGCATCGATGAAGAACGCAGCGAAATGCGATACG-
TAATGTGAATTGCAGAATTCAGTGAATCATCGAGTCTTTGAACGCACATTGCGCCCTCTGGTATTCCGGAGGG-
CATGCCTGTCCGAGCGTCATTGCTGCCCTCAAGCACGGCTTGTGTGTTGGGCTCCGTCCT-
CCTTCCGGGGGACGGGCCCGAAAGGCAGCGGCGGCACCGCGTCCGGTCCTCAAGCGTATGGGGCTTTGT-
CACCCGCTTTGTAGGACTGGCCGGCGCCTGCCGATCAACCAAACTTTTTTCCAGGTTGACCTCGGATCAGGTAGGGA-
TACCCGCTGAACTTAAGCATATCAATAAGCGGAGGAAA
```

The sequence was pasted into NCBI BLAST and was found to be an exact match for several sequences of the species *P. brevicompactum*.

PHYLOGENETICS

The targeted barcode mold sequenced in this study matched several different sequences of the *P. brevicompactum* species in the GenBank database. The phylogenetic tree graphs four *P. brevicompactum* in relation to the targeted mold. The closest relatives to the identified fungi are all of the *Penicillium* genus, including species *Penicillium biourgeianum*, *Penicillium simile*, and *Penicillium mexicanum*. Other distant relatives include *Penicillium rubens*, used as an antibiotic, and *Penicillium quebecense*.

To read a basic phylogenetic tree, two attributes are considered: branches and nodes. Branches are horizontal lines that represent evolutionary lineage changing over time. Nodes are points on the line and can be endpoints (external nodes) or intersections (internal nodes). An external node represents an organism that has been sequenced used as an endpoint marker. An internal node represents a shared ancestor of a lineage before diverging. The number next to an internal node is the probability of support for that node.

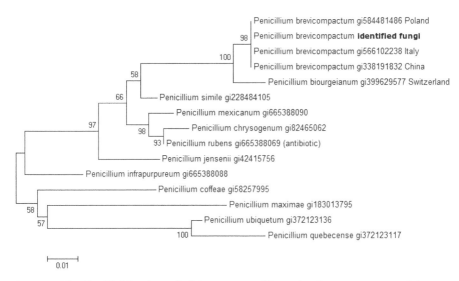

Figure 2-3. The identified fungi matched the sequence of four P. brevicompactum around the world. Analysis of the tree predicts that P. brevicompactum had a common ancestor with P. biourgeianum before splitting into different species. From thousands of Penicillium sequences in GenBank, a few were selected for the tree.

Discussion

This experiment reiterates a basic procedure to isolate and identify an unfamiliar organism. The streaking and subculture method may be used to isolate most microorganisms. Microscopic identification can be a practical approach to discern cell type. Barcoding DNA, a relatively novel approach, is a very cost-effective way to determine an organism's species without sequencing entirely. The phylogenetic tree relates isolates to similar biological species.

P. brevicompactum can be found on temperate forest soil and moldy fruits. It is unknown if *P. brevicompactum* is the primary mold infection of the *C. limon* fruits, and further experimentation is needed to come to a conclusion. Results conclude that *P. brevicompactum* is naturally growing on *C. limon*, affecting its growth.

With the mold species isolated and identified, the target mold, *P. brevicompactum*, may be used for other novel experiments. Novel protocols based on bacteriophage filtration could be used to collect mycophages, which are phages specifically targeting fungus, in the soil, and these could be applied to plaque assays to see if they counteract the invasive mold. The use of mycophage filtration may be a safe and cheap alternative to collecting mycophages to use as a pesticide.

Industrial farms frequently spray chemical pesticides over the *C. limon* crop, often at the expense of the health of workers who are exposed to large amounts of pesticides.[6] Using mycophages may be a safe alternative. The use of phages on poultry products has already been approved in the United States by the Food and Drug Administration to prevent *Salmonella* infections.[7]

References

1. Robert A. Samson and John Pitt (eds.), 2000. *Integration of Modern Taxonomic Methods For Penicillium and Aspergillus Classification.* CRC Press. 14-15.

2. Kenneth B. Raper and Charles Thom, 1949, *A Manual of the Penicillia* (Baltimore, MD: The Williams and Wilkins Co., 1949).

3. John Pitt, *The genus Penicillium and its teleomorphic states Eupenicillium and Talaromyces* (London: Academic Press, 1979).

4. Jens C. Frisvad and Robert A. Samson, "Polyphasic taxonomy of Penicillium subgenus Penicillium. A guide to identification of food and air-borne terverticillate Penicillia and their mycotoxins," *Studies in Mycology.* 49 (2004): 1-174.

5. Fungal Barcoding Consortium, "Nuclear ribosomal internal transcribed spacer (ITS) region as a universal DNA barcode marker for Fungi" *PNAS* 109, no. 16 (2012): 6241-6246.

6. Linda A. McCauley et al., "Studying Health Outcomes in Farmworker Populations Exposed to Pesticides," *Environmental Health Perspectives* 114, no. 3 (2006): 953-960.

7. The Food and Drug Administration: Agency Response Letter GRAS Notice No. GRN 000468.

Louis Huang recently returned to the US after tracking Pecari tajacu in the wetlands of Brazil for the WCS. He previously worked as a web developer for Virgin America and Stitcher (now Deezer Talk) in San Francisco. He holds degrees in chemical engineering and economics from Syracuse University and has done genomics coursework at Harvard University.

Alan Rockefeller specializes in mushroom taxonomy. He has been collecting mushrooms for 10 years and has traveled to Mexico to collect mushrooms for the past 8 years. Alan is a network security expert, as well as a moderator at the Shroomery Mushroom Hunting and Identification Forum, and posts all of his mushroom photographs on mushroomobserver.org. When he is not hunting mushrooms Alan spends his time looking at mushrooms under the microscope, soldering electronics, hacking Unix and sequencing mushroom DNA.

From Student Protest to DNA Synthesizer

Alexander Murer

Roughly five years ago, studying molecular biology in Graz, Austria had become routine for me, even though I just had started two years before. My dreams of an open, exciting, and less restrictive chapter in my educational career had been crushed between the authoritarian university system and its routine of boring, one-way lectures, memorization of books, and writing tests.

Luckily, everything changed one night when I met a friend at a party who told me students would be gathering to protest new fees, the elimination of voting rights for student representatives, entrance exams, and increased pressure to finish a degree, which they believed led to educational negligence. I went to the protest to see what was going to happen and found myself at a lecture-hall occupation a few hours later. Those three months of occupation turned out to be the most intense experience of my educational career. I was encouraged by my fellow protesting students to think about education and speak up, to vote at the daily meetings and take part in change, not just consume.

We developed four core demands:

Democratization
> Universities in Austria are public institutions, but the greatest percentage of voting-age people there—the students—have few to no voting rights.

No acceptance tests
> There's no serious way to determine whether someone is going to be good in a particular field of science before letting them actually learn about it.

No fees
> How can you ask for fees from students, who usually don't have much money?

No time pressure

Science, education, and engineering take time, both to study and to benefit from experience at outside institutions.

After three months, the protests and occupations ended, but my frustration did not. It was not only the university's laws and framework that annoyed me, but also the teaching style. Instead of giving students hands-on experience and encouraging self-reliant experiment-based learning, professors make them memorize entire books—especially the first-years—and take theory tests, which impede fast progress. I believe that students should have free access to a lab right away and learn the theory while experimenting. These thoughts motivated me to look for a way to do my own projects in the field of biotechnology.

One day I stumbled across pictures of a bioreactor in a university script. I immediately decided that it couldn't be that hard to build my own and threw the script into the corner. A pretty long, enlightening, and sometimes painful journey had started. While my first attempts were rather basic, two friends soon joined the project: Bernhard Tittelbach, who is our hardware programmer and electronics engineer; and Martin Jost, a software programmer and classmate of mine in molecular biology. We tinkered together at the local computer hackerspace "realraum," which offered room, a great tinkering atmosphere, and some tools.

Amazed by the huge potential of community workshops like "realraum," we soon made another important move: I saw that we needed another community workshop or hackerspace, but for molecular biology. Just a few months later, in January 2013, Martin (along with everyone else who showed interest in the project) and I founded the first biohackerspace in Graz: Open Biolab Graz Austria, or simply OLGA. We were pretty lucky that friends had a spare room in their workshop and agreed to rent it to us cheap.

With our biohackerspace, our ambitions to develop a bioreactor became more serious. We soon founded our first company for the reactor, named Briefcase Biotec, to get a small grant for hardware costs.

But the bioreactor was just the beginning; once you start to see potential everywhere and do something about it, you find yourself flooded with new ideas. Developing a DNA synthesizer is somewhat the holy grail of the DIYbio community, so obviously the topic sparked our interest. Looking for some funds, we came across the Synbio Axlr8r program (now Indie.bio) by SOSVentures. During the following great months in Ireland, our Kilobaser, a.k.a the Rapid DNA Prototyper, was born.

DNA Synthesis

The artificial synthesis of DNA was pioneered in the early 1950s. Since then, several chemical approaches have been developed, but the most commonly used method today is nucleoside phosphoramidite synthesis, which was introduced in 1981. Even though it is the backbone of each and every development in the field of genetic engineering, DNA synthesis still suffers from a series of problems:

- While prices for sequencing DNA have dropped dramatically, prices for synthesizing have stayed almost the same.

- The maximum length you can synthesize in one go is only about 200 base pairs, since even in optimized systems there's a 0.5% chance of an error for each base added.
- Longer DNA constructs are built from several shorter strands, which makes synthesizing an error-prone and labor-intensive process.

The issue we focus on is that today almost all labs rely on outsourced synthesizing services, leading to long turnaround times, loss of data security, and high costs in the long run. There are more than 100,000 life-science labs worldwide, but only about 250 synthesizers sold each year. Why would all these labs rely on others rather than using their own devices? It turned out to be perceived as much more convenient to order your DNA online, even though you'd have to deal with ever-accumulating delays, than to run a complicated machine yourself and have to deal with more than a dozen different and difficult chemicals in addition to several manual steps required to finish the synthesis.

So when describing our invention, I usually start by talking about the beauty of desktop printers and coffee machines. Who would have thought that something as simple as coffee packed in a cartridge would have such a heavy impact on the coffee industry? It's the user experience that determines whether a product turns out to be a huge success or not. Yeah, we are developing cartridge-based DNA synthesizers. You just enter your sequence and hit the synthesize button. Our aim is to bring DNA synthesis back to the lab bench, where it belongs. But we are not very ambitious nucleoside phosphoramidites if each life-science lab has just one Kilobaser, we'll be happy!

Alexander Murer is the founder of Open Biolab Graz Austria and Briefcase Biotec. His attendance at student protests in 2009, where he adopted ideas of a more self-reliant, open, and free education, ultimately led him to independently work on biotechnology-related projects and to found his start-up company Briefcase Biotec. Later, he founded the biohackerspace, Open Biolab Graz Austria, which is now the first biohackerspace in Europe to hold a GMO license. His recent project Kilobaser is a cartridge-driven, rapid DNA prototyping machine.

Open qPCR: Open Source Real-Time PCR and DNA Diagnostics

Josh Perfetto

In November 2014, we launched a Kickstarter campaign for Open qPCR, which raised \$202,701 and dramatically lowered the cost of a general-purpose real-time polymerase chain reaction instrument (such devices typically sell for more than \$20,000). More importantly, we just open sourced a technology that is the gold standard for diagnosing viruses in humans, detecting pathogens like *Salmonella* in food, genotyping many genetic variants, and a myriad other applications. While the first Open qPCRs are being manufactured this winter, I wanted to take a moment to explain real-time PCR and its benefits.

Let's start with basic PCR. Suppose we want to know if a patient is infected with malaria. Detecting the malaria parasite in a patient's blood is like finding a needle in a haystack. To do so, we first need to amplify the malarial signal above the background noise, which is exactly what PCR does.

All of the DNA extracted from the patient sample goes into the reaction, and only a specific target DNA sequence is selectively copied. PCR is a chain reaction of typically 30–40 cycles, and each cycle of the reaction doubles the amount of target DNA present. Thus, if malaria DNA is targeted PCR exponentially amplifies this DNA to detectable levels, but *if and only if* the malaria DNA was present to begin with. Facilitating this chain reaction requires the repetitive heating and cooling of the reagent/DNA mixture, which is what a basic "endpoint" thermocycler such as OpenPCR does. However, with these endpoint machines, you're left with a tube of DNA as your output; you don't know whether the patient has malaria without performing additional laboratory work.

Real-time PCR takes things one step further by integrating an optical fluorescence detection system into the instrument. This lets the instrument detect the amplified DNA, and thus provides information rather than DNA as an output. In a process executed by a single device, you can determine whether the patient has malaria and see the result on a computer interface.

At this point, the astute reader of *BioCoder* may be saying, "Aha, but I could use a basic thermocycler and then run a gel!" To an extent you could, but if you went to your doctors and had a blood test run for a suspected virus or pathogen, you could be virtually assured that the lab was not running a gel. Let's examine the benefits of real-time PCR and why it's often considered to be the gold standard for diagnostics.

Benefit #1: More Specific Detection Reduces False Positives

The specificity of the basic PCR reaction depends on two DNA primers annealing with their complementary sequence. However, such primers can frequently "misprime" and anneal elsewhere, amplifying an unintended sequence of DNA. Many real-time PCR diagnostics utilize an additional fluorescently labeled "probe" DNA that must anneal to a specific sequence within the amplified segment for a signal to be detected. In the event that the two PCR primers misprimed, it is highly unlikely the probe would find a complementary sequence within the incorrectly amplified sequence. Other real-time PCR tests don't use probes, but measure the melting temperature of the amplified sequence to ensure it is the correct one. Both techniques help avoid false-positive results.

Benefit #2: Internal Positive Controls Reduce False Negatives

In any PCR test, you need to run a positive control: if the positive control doesn't amplify, you can't interpret a sample's failure to amplify as a negative result. However, with standard PCR, this positive control is run in a different tube, meaning it's possible for the control reaction to work but the diagnostic reaction to fail for some other reason (pipetting error, PCR inhibition, etc.). With real-time PCR, the control and diagnostic reactions can occur in the same PCR tube, and the results can be differentiated using different optical channels. This is known as an internal positive control, and it eliminates a number of problems that can cause false negatives.

Benefit #3: Quantitative PCR (qPCR)

In many applications, such as analyzing gene expression or measuring viral load, it is necessary to not only detect but also quantify the amount of DNA present. With endpoint PCR, you make a fluorescence measurement only after the entire PCR run is complete, and you typically detect either a whole lot of DNA or none at all.

With real-time PCR, fluorescence measurements are made after each PCR cycle. During the first several cycles, no signal is observed, but eventually a weak, then strong, and then saturated signal will be seen. A real-time instrument uses not only the level of signal observed but also the cycle number of when it was observed, to quantify the amount of DNA present (thus the term qPCR). For example, suppose a certain signal level of sample A is seen at cycle 15, while the same level is not seen in sample B until cycle 18. Assuming 100% efficiency, you could conclude that sample A contained eight times the starting DNA as sample B, as the target DNA present doubles each cycle (in practice, you would measure the efficiency rather than assume it, and the instrument would employ a much more sophisticated calculation).

Benefit #4: Saves Time and Money

A standard endpoint thermocycler takes in DNA and gives out DNA. Obtaining a diagnostic result from that output DNA requires further downstream processing, such as casting and running a gel. Doing so requires time and reagents, which of course equals money. Once you have a real-time PCR thermocycler, the per-sample cost is actually quite cheap, and you can get diagnostic results in as little as 30 minutes.

Benefit #5: Genotyping

A common use of real-time PCR is for genotyping, or determining the specific allele of a gene present. PCR primers can commonly be designed to amplify a region of DNA that contains the variants of interest. But once it's amplified, how can you identify the internal sequence?

In genotyping applications, there are commonly a small number of potential sequences present, and these sequences will have different melting temperatures (the temperature at which the double-stranded DNA dissociates into two single strands). A real-time PCR instrument can detect which variant is present by doing a melt-curve analysis: the temperature is slowly increased while the fluorescent signal is monitored. As the fluorescent signal is proportional to the amount of unmelted double-stranded DNA present, the melting temperature can be determined.

Open qPCR

We've talked of the many advantages of real-time PCR but neglected the typical disadvantage: cost. Real-time instruments typically cost $25,000 to $50,000, which greatly restricts access to this technology. With Open qPCR, we set out to reduce the cost by an order of magnitude and deliver a modern web-based interface that makes the device easy to use.

Open qPCR is a real-time PCR thermocycler that processes 16 samples at a time and ramps the temperature at 5 degrees Celsius. Such fast ramping speeds enable fast run times: test results can be obtained in as little as 30 minutes. The device employs a solid-state optical detection system consisting of LEDs, photodiodes, and optical filters. Open qPCR is available in both a lower-cost, single-channel version, and a dual-channel version that can detect multiple fluorophores at the same time and thus utilize internal positive controls.

Open qPCR is built upon the Beaglebone Black development platform, which runs embedded Linux on a 1 GHz ARM processor. The device has Ethernet, WiFi, and USB connectivity. We also included an embedded touchscreen interface, so the device may be used in the field without any PC connectivity.

Ease of use depends a lot on the software, so we spent a lot of time on user-interface design. Open qPCR runs an internal web server to deliver an HTML5/JavaScript interface, which can be accessed via any computer or smartphone. This software allows the user to graphically edit the PCR protocol; define the plate layout; and view amplification curves, standard curves, and interpreted diagnostic results. An open REST API allows Open qPCR to be controlled by external automation systems.

Open qPCR is also completely open source: the mechanical and electrical CAD designs, real-time control software, scientific analysis software, and web interface will all be released as open source when the machine ships in March 2015.

Our eventual goal is to make reliable DNA diagnostics available not only in the lab, but to all who have innovative ideas on how to apply this technology: beekeepers with a theory on population decline, consumer activists fighting fraudulently labeled food, and educators teaching a new generation of bioliterate students. The launch of Open qPCR is a first step, but there's far more to come to complete this vision.

Josh Perfetto is the founder of Chai Biotechnologies, a Santa Clara, California–based start-up creating modern tools for molecular and synthetic biology and deploying the power of these technologies outside of the lab. Prior to founding Chai, Josh led engineering teams in a number of software and wireless start-ups, and in 2010 he cofounded OpenPCR, creators of the world's first open source PCR thermocycler. He can be reached at josh@chaibio.com or @jperfetto. Learn more about Open qPCR at www.chaibio.com/openqpcr.

Other Selves: An Artistic Study of the Human Microbiome

Joana Ricou

Abstract

We are made of many parts and many types of parts. Many of these are alive, and as the study of the human microbiome has recently revealed, most are not human. I explore the microbiome as an artistic medium and the questions it raises about identity, since the microbiome blurs the boundary between organism and environment, figure and ground. I cultivated my microbiome and others inhabiting my environment to create living paints with which to make an unusual self portrait: an otherself portrait. In the process, I arrived at a series of otherself landscapes, discovering the body as a world, multiple and dynamic.

Background

Comparing the concepts of identity in art and biology yields some insights about how our understanding of ourselves evolved. In art history, we can observe the progression from a concern with stylized forms and ideas, to a true depiction of reality, to an exploration of the metaphorical and subjective. Meanwhile, identity as defined in biology mostly limited itself to polemic, taxonomic descriptions of the human species[1] until the 20th century, when it exploded to the forefront of the conversation with the discovery of DNA and the uniqueness of the genetic finger-print.

More recently, several discoveries in biology, such as the microbiome, have fundamentally challenged our notions of identity. Researchers determined that not only are nonhuman organisms essential to our health, but nonhuman cells outnumber human cells 10:1, and nonhuman genes outnumber human DNA

150:1.[2] Science now proposes that we consider the human being not as a single organism, but as a diverse ecosystem.

I am interested in the way that biology defines and redefines boundaries in the human body, and I use these discontinuities as starting points to phrase questions about identity. I was intrigued by the idea that there is a nonhuman part to each of our many parts, most parts. And since our microbiome stems from our environment—the air we breathe, the food we eat, the objects we touch—it connects us to our environment directly and blurs the most fundamental boundary of individual identity: skin.

The Microbiome as an Artistic Medium

I consulted with scientists to determine a protocol to cultivate microbes commonly found on the surface of the human body and in an urban environment. Using Q-tips, I collected samples of my microbiome, swabbing only external surfaces of my body (face, scalp, and hands) and my environment (air, table, and floor). The collection was done under semisterile conditions, and each sample was plated in a petri dish with LB agar and potato dextrose, then incubated at near body temperature.

I used the standard crosshatching pattern for plating so the initial petri dishes didn't differ visually from common scientific sampling of bacteria. To the untrained eye, the diversity of the microbes grown was exciting: there were a variety of colors, ranging from whites to yellows to reddish orange, and a variety of textures, many bacteria and some fungi. I had hoped that samples grown from different parts of the body would look different, since our microbiome is thought to vary by nook and cranny, but they were not recognizably different. Consulting with experts, I learned that the environment of the human body is truly unique and the in vitro conditions cannot match its richness. My own inexperience may have limited the diversity; for example, the temperature of incubation was likely too high because the cultures were taken from the surface of the body, which has a slightly lower temperature than the interior (body temperature). In effect, I had selected for a hardy few. I digitally combined all the human samples in one image, virtually integrating organisms that couldn't tolerate each other in vitro and creating a more complete portrait.

Whereas there was no significant difference between the different human microbiome plates, there was almost no overlap between the human and environmental samples. This was disappointing, since I had hoped to see the similarity, the blurring between organism and environment. Comparing a combined image of all human- with all nonhuman-derived microbes, the two works have a distinct cosmic feel: individual dots and tendrils float in the environment, the petri dish boundary creating a planetlike shape and emphasizing the loss of connection between the two.

Individual bacterial colonies and samples of different fungi were carefully picked off each plate and grown separately, in liquid LB agar, with agitation, at body temperature. These conditions allowed for rapid growth and the creation of a palette of living colors.

An Unusual Self-Portrait

The cultures were used as common paints, as I painted two compositions on a series of petri dishes using traditional brushes (Figure 5-1). The petri dishes contained LB agar. One composition was a self portrait, a bust. Cultures from the forehead or chin were used to paint those areas of the face; cultures from the environment were used for the background. The second composition was symbolic, a filled circle surrounded by a ring, symbolizing the person and the environment, respectively. In this case, human paints were mixed in the center, and environmental paints were used in the ring.

Figure 5-1. Creating a palette of living colors. The "paints" included bacteria and fungi derived from the collected microbiomes of myself and my environment.

Over three days, all of the plates in this experiment revealed a similar result: a milky white, homogeneous substance, likely due to the artificial decisions made about the growth conditions. The resulting work was an unusual self portrait (Figure 5-2), an otherself portrait, which had an interesting mix of markmaking between those made by hand with brushes and the textures and forms created by the microbes. Between the two, the outline of an individual remains barely discernible, effectively blurring the boundary between the figure and the ground.

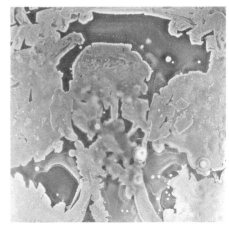

Figure 5-2. Joana Ricou, Me and My Other Selves, inkjet, each 22 × 22 in., 2014.

Over three days, all of the plates in this experiment revealed a similar result: a milky white, homogeneous substance likely due to the artificial decisions done about the growth conditions. The resulting work was an unusual selfportrait (Fig. 2), an otherself portrait, which had an interesting mix of markmaking between those made by hand with brushes and the textures and forms created by the microbes. Between the two, the outline of an individual remains barely discernible, effectively blurring the boundary between the figure and the ground.

Exploring the portrait at a higher resolution revealed a new layer of detail. With the aid of photographer Raúl Valverde, I discovered that the seemingly homogeneous biofilm had a surprising variety of shapes and textures. Beautiful and delicate borders arose from complex interactions stemming from physical, chemical, and biological tensions created by the microbiome.

At this degree of magnification, the imagery became topographical, and the reference framework switched from portraiture to landscape representation (Figure 5-3). Searching for a portrait of my other selves, I feel that I discovered a selfhood as a world.

Figure 5-3. Joana Ricou, Other Landscapes (Microbiome) no.1, 35 × 55 in., 2014 (photographer: Raúl Valverde).

Conclusions and Next Steps

I started this project looking for a connection between the body and the world and felt that I arrived more at a discovery of the body as a world. Continuing to discuss these results, I learned that a significant part of our microbiome comes not from our world but from our mothers (from vaginal delivery). I am expanding the project now to include how the microbiome blurs not just the boundary between the organism and the environment and but also between generations. From a technical perspective, the microbiome is surprising in many ways: control has to be shared with a living medium but, like with any other medium, manipulating the conditions of its use seems to lead to interesting results.

Biology has uncovered the body as multiple, dynamic, and transient, and the microbiome is only one of the latest research areas to embody these themes. I believe that an artistic exploration of our biology is not only fun but essential, not necessarily for helping us understand what our biology is or what it does, but what it means. Playing with portraiture and apparently landscape, we can explore and incorporate these changing ideas into the changing image we somehow insist on recognizing every day in the mirror.

This project was done at the Bioart Residency at the School of Visual Arts in New York City, with mentorship from Suzanne Anker and Brandon Ballengée. Thanks also to Sebastian Cocioba, Rául Gomez Valverde, Dr. Jerry Schatten, Dr. Stephanie Davis, Dr. Martin Blaser, and Dr. Maria Dominguez Bello.

References and Notes

1. Mark Jarzombek, "Are We Homo sapiens Yet?" *Thresholds 42: Human Journal of the MIT Department of Architecture* (2014): 10-17.

2. B. Zhu, X. Wang, and L. Li, "Human gut microbiome: the second genome of human body," *Protein Cell*, no. 8 (2010): 718-725.

3. David A. Relman, "Learning about who we are," *Nature* 486, no. 7402 (2012): 194-195.

Joana Ricou works at the intersection of art and science as an artist and as a creative consultant in education. She has collaborated and exhibited at galleries, museums, and universities including the Andy Warhol Museum, Carnegie Science Center, New York Hall of Science, Harvard University, and Children's Museum of Pittsburgh, among others. Joana has a Bachelor of Science and Arts from Carnegie Mellon University (2004) and an MS in Multimedia Arts from Duquesne University (2009).

iGEM's First Giant Jamboree

Edward Perello

I expect that BioCoders are familiar enough with synthetic biology to know (and love) what iGEM is all about. I was once like you. I had been to the local events and regional jamborees, and I thought they were great. But I had never been to the Boston Jamboree, and the 2014 competition was unlike any other to date.

A Giant Jamboree

For a few days every year, Boston becomes home to some of the most impressive, outlandish, and curious people and projects in the synthetic biology domain and its supporting fields. This year, to celebrate the competition's 10th anniversary, iGEM HQ invited *all* participating teams to join together at the same time in one Giant Jamboree (*http://2014.igem.org/Giant_Jamboree*).

At the five-day event, 245 teams (*http://bit.ly/bio6-teamlist*) gathered under one roof to demonstrate new organisms, concepts, data, and protocols to one another and the who's who of synbio, including 105 judges (*http://bit.ly/bio6-judges*) and hundreds of other experts from all over the world.

Generally speaking, teams pursue technically challenging projects using feasible (though oftentimes high-risk) biological approaches, and so projects explore a vast array of moonshots to change the way we manufacture materials, treat diseases, generate energy, or create artistic masterpieces using living organisms and natural materials.

And though iGEM is a competition only in title (there are no points), teams typically devote their waking lives to their work. Accordingly, medals and other awards are presented to the teams that merit them. But for most participants, the ideal outcome is a successful proof of concept that demonstrates the validity of a novel synthetic biology approach that enables future work by other iGEM teams or external groups. Many iGEM alumni even look to spin out companies, like Bricobio's Hyasynth Bio (*http://hyasynthbio.com/*) in Montreal, UCL's Bento Bioworks (*http://www.bento.bio/*) in London, and UCC-Ireland's Benthic Labs (*http://bit.ly/bio6-benthic/*) in Cork.

A Few New Projects and Technologies

The quality of projects at every turn was outstanding, and it is simply an injustice (though also a necessity) to mention just a few to keep the article short. I therefore apologize to all those I skip, and look to name a few that got me thinking.

SDU-Denmark, on the food and nutrition track, challenged my culinary pre-conceptions by generating edible *E. coli* (*http://bit.ly/bio6-ecoli*), which had been modified to secrete nutritional proteins, essential fatty acids, and a palatable limonene from limonene synthase. Unfortunately, I wasn't able to try it, and still I wonder how they will market this product, but I'd certainly give it a go.

Utah State, competing in the manufacturing track, helped me clean clothes by functionalizing plant-degrading enzymes onto a bioplastic bead. They brought it all together as a bona-fide "stain-buster" laundry product (*http://bit.ly/bio6-stain*) that can be added to a wash.

I was also pleasantly surprised to see that so many teams embraced CRISPR genome-editing technologies as part of their projects. I counted at least two dozen projects that used CRISPR to directly edit a target gene to achieve a desired outcome, or alternatively made use of gRNAs for sequence-specific targeting of pathogens, such as CU-Boulder's Pathogen Assassin (*http://bit.ly/bio6-pathogen*) and Uppsala's Bactissile (*http://bit.ly/bio6-bactissile*) system.

On the hardware side, Cooper Union (*http://bit.ly/bio6-dnasynth*) brought together some beautiful engineering and novel biochemistry to deliver a prototype of a template-free DNA synthesizer. This is quite an achievement, given that the college has only recently forayed into biology with the opening of the Medvedik lab last September. As for software, USTC-Software built Biopano (*http://bit.ly/bio6-biopano*), a gorgeous (and very useful) online tool for visualizing gene regulation networks and metabolic relationships—all for future iGEM teams and other synthetic biologists to use. I'd really encourage everyone to take a look at the software projects, as they can be downloaded and used for free.

A Few New Tracks

In fact, the software track was one of many new tracks debuting at the Giant Jamboree. The majority of these were simply dedicated subcompetitions for areas of research that had been part of old-school tracks in earlier competitions. In addition to software, policy and practices, measurement, and art and design all had their own judging criteria and awards systems.

For most reading *BioCoder*, the most pressing development was that, for the first time, a specific track for the DIYbio community was put together. In the past, iGEM HQ's refusal to allow biohackers to compete as standalone teams amongst core academic teams had been a point of contention for many in the community. However, this attitude has softened, to the point where six new teams came to the Jamboree as part of the new Community Labs track.

New York's Genspace demonstrated its Open Lab Project (*http://bit.ly/bio6-genspace*), a complete set of knowledge, tools, and resources required to success-fully develop a thriving community biolab, following the open source principle of encouraging more community labs to form worldwide. Baltimore's BUGSS expanded on its 2010 work to develop affordable polymerases (*http://bit.ly/bio6-polymerase*) for the community by developing a kit for the easy and cost-efficient purification of Pfu polymerase, as well as an *E. coli* strain to sequester heavy metals.

California was strongly represented in the community-labs track, with a combined San Francisco Bay Area team (Biocurious and Counter Culture Labs) bringing along the now-famous Real Vegan Cheese (*http://bit.ly/bio6-cheese*) project, to great fanfare. The LA Biohackers (*http://bit.ly/bio6-genome*) group went all out and attempted to boot up *Streptococcus thermophiles* in *Bacillus subtilis*, and then use cre/lox recombination to remove the *B. subtilis* genome. Though they ran out of time before the Jamboree, this is some seriously cool work for a bio-hacker group to be undertaking, and its scope genuinely impressed me.

My personal favorite was the arguably simpler, yet clearly gorgeous bacterial e-pixel system (*http://bit.ly/bio6-epixel*) brought over from San Jose by the Tech Museum of Innovation team. Their stochastic light-emission vectors were abso-lutely stunning, to the point where their organisms are a clear contender for awards at Burning Man 2015, let alone iGEM 2014.

London BioHackspace (*http://bit.ly/bio6-juicyprint*) was the only non-US team on the track, and they brought along samples of their lab-grown bacterial cel-lulose from a prototype "JuicyPrint" system—a 3D printer that can be fed with fruit juice. I was excited to see that a European team was represented amongst the US-heavy community-labs track, and hope to see a lot more compete next year.

Having said that, it will be interesting to see how the DIYbio community con-tinues to engage with iGEM in the future, and whether or not participation by these labs grows. I received a lot of comments from individuals in the DIYbio community that the high costs involved in iGEM were a major barrier to participa-tion, to the point where they may not return in 2015. This is a point of distaste shared by many iGEM participants; I've heard rumors that even well-funded uni-versities can only send a team once every two years.

New People, New Skills, New Competitions?

It seems that the decision to add new tracks continues to open up the iGEM competition to a much wider range of participants. For instance, in a survey of software-track participants, I found that over 50% were software developers with limited experience in biology, bioinformatics, or computational biology. That is to say that before the competition, over half of the team members had never been in a lab or picked up a biology book.

This is quite an exciting find, because it suggests that synthetic biology is now rapidly filtering into a wider range of fields *and* pulling in new people. Though my evidence is, at this stage, based on limited data, I'm excited to see a synthetic biology competition foster cooperation between biologists and non-life-science professionals in this way. In order to confirm this expectation, I would encourage others to conduct further assessments of the skills represented within teams across different tracks. If we are able to understand the makeup of teams, we can ensure that iGEM HQ has sufficient understanding to support the new tracks as they grow.

On the other hand, I also wonder if these new tracks can remain part of iGEM forever. The software track certainly has a life of its own, and we will have to see whether or not community labs will embrace iGEM or go in a different direction. We may very well see new competitions, based on the iGEM model, form for each new community, in the same way that BIOMOD (*http://biomod.net/*) has kicked off for the DNA origami community. Alternatively, these perceived teething problems may simply be a product of every track at iGEM having its own culture. iGEM has a great history of improving itself year after year, and I am certain ways will be found to accommodate the needs of the growing community.

Excitement at Every Turn: Breakout Sessions

Despite their differences, all tracks are unified by their shared energy, excitement, and thirst for scientific endeavor. Across three floors and dozens of rooms, I saw in peoples' eyes the same super-excited look that my little nephew gets when he shows me a new Lego model. At the Giant Jamboree I saw 2,300 people who had that look, except that their Lego sets happened to be made from standard biological parts, could self-replicate, or could otherwise help you design something that can.

With my hand on my heart, I can say that the iGEM poster sessions happening between team presentations simply could not be beaten. Seeing them alone would have been enough to leave me feeling more than satisfied with the experience.

But the Giant Jamboree just kept on delivering, with breakout sessions to suit everyone's tastes. The Outreach Workshop on Biodesign Automation brought together the best in automated design of CRISPR gRNAs, genome-editing vectors, and DNA-assembly protocols, with demos from BBN Technologies (*http://www.bbn.com/*) and Desktop Genetics (*http://www.deskgen.com/*). The Cultured Products showcase, hosted by Ginkgo Bioworks (*http://ginkgobioworks.com/*), featured talks by artists creating biofabricated clothing, scientists uncovering the design principles of fermented foods, and iGEM alums who had designed yeast that smells like roses. The entrepreneurship panel discussion really surprised me, as it was *completely* rammed with students eager to find out what it takes to build a successful synthetic biology start-up. This is very encouraging, and I'm excited to see who might resurface at SynBioBeta (*http://synbiobeta.com/*) next year.

There were also numerous talks about policy and practices. In addition to the FBI's annual outreach program and keynote, the EU-funded Synenergene Project (*http://www.synenergene.eu/*) ran talks almost every day. This relatively new NGO had provided financial assistance to eight iGEM teams in 2014, and in each session, the iGEM teams talked about how they had imagined future challenges for humanity, and then designed their projects around those challenges to explore not only the technical solution to their problem, but the moral and philosophical framework surrounding it. Given the somewhat science-fiction vibe that iGEM projects can give off to non-scientists, I think this is a fantastic way to get students to think critically about the point of their technology and its role in society. At the very least, it equips them with the skills to parlay with the public on the need for certain areas of research, a necessary talent for any synthetic biologist.

One thing that I felt was missing was the mainstream media. I concede that repressilators and quorum sensing may perhaps bore most people, but the application of biology, and certainly the technical approaches taken by the teams, are inspired enough to warrant more attention. At the same time, I'm not too upset about this and expect that the media's presence is strictly controlled by iGEM HQ. I also noticed quite a few sniffer dogs. I'm unable to say whether or not this is a standard feature for the Hynes Convention Center or simply a measure to safeguard the 2,300 genetic engineers gathering in the same place.

Get Yourself to iGEM 2015

If you didn't have a chance to attend the iGEM 2014 Giant Jamboree, then I urge you to get involved. If you are a student, stop reading and start a team as soon as you can; as we go to press, the time to start making this happen has arrived! If you are an academic thinking about going, inspire your students and challenge them to push some biological boundaries. If you are part of a community lab, participating is not only a great opportunity to connect with academic groups, but also a fantastic way to expand your parts repositories and networks.

To everyone reading this: if you need support in your DNA assembly, cloning, and CRISPR workflows (especially if you spend hours running cloning parties and planning assemblies), or simply want a tool to manage your iGEM repository, year after year, for free, get in touch with us at Desktop Genetics. We're now enrolling over two dozen iGEM teams to make full use of our AutoClone and CRISPR tools and expanding our support program to all teams in 2015. We can save you a lot of time and money, and I'll be elaborating on how in a future article.

I will certainly be going back to the Jamboree next year. The explosion in ingenuity, and the feeling of genuine progress—all at the scale afforded by a Giant Jamboree—was a second-to-none affirmation of where we are headed in this field. I hope iGEM HQ sticks with the model, and I hope to see you all there, where we can collectively have our minds blown in Boston.

Edward Perello is a founder and director of business at Desktop Genetics, a London-based software company that develops tools for genome editing, synthetic biology, and cell-line engineering. He can be contacted at edwardp@deskgen.com. Desktop Genetics is recruiting talented developers to join full-time in London.

A Simple Data Acquisition and Plotting System for Low-Cost Experimentation

Jonathan Cline

Abstract

This paper details the design and construction of a simple data acquisition and plotting system for amateur or commercial use. The system uses a general-purpose computer; a standardized, USB-powered peripheral board; and a small software application to measure, record, and plot a real-time signal. The system is demonstrated in use with a biochemistry chromatography experiment; however, the system may be used in conjunction with any experiment that needs to measure and plot an electrical signal. Modifications to the hardware to adapt the circuit to various signal voltages are discussed. Total cost is about $35 in parts, a few hours connecting common electronic components with a soldering iron, and a few hours configuring software on any typical computer. This system does require a separate computer that supports USB, although you can substitute a small, inexpensive embedded computer, such as the popular Raspberry Pi board (currently $25). In contrast, a commercial-grade chart recorder that performs the same functions might have a price tag of $2,000. Building this basic data acquisition system also yields the understanding of how to design such systems and use them either in laboratory protocols or in field experiments.

This paper follows a typical engineering format: laboratory requirements, design considerations, theory of operation, and build instructions. Suggestions are included for improving the design. Further discussion is welcome on the DIYbio discussion group (*http://bit.ly/bio6-diybio-forum*) or via email (*jcline@ieee.org*).

Background and Theory of Operation

The impetus for this system was the desire to perform a particular experiment in a DIYbio lab. The experimental protocol called for separating a biochemistry solution into two parts: the desired target macromolecule and the waste. A chromatography apparatus was chosen to accomplish this separation (explained below, and shown in Figure 7-1). The chromatography apparatus generates an electrical signal (an *analog voltage*) while the experiment is running, corresponding to the concentration of the desired target. When a certain voltage is reached, the measurement indicates that the biologist can proceed with the next step of the protocol, extracting a high concentration of the desired target from the solution. The success of our experiment depended on a precise chromatography measurement (the signal voltage) and visually presenting this measurement versus time while the experiment was run.

In general, biochemistry experiments are performed in solution or liquid culture. A critical step is extracting the desired target molecule or cellular body from the suspension. This requires a method of measurement that can discriminate the desired target from the background. Commonly, this is performed with *spectrophotometry*. A spectrophotometer has two parts: an emitter and a detector. If the *emitter* produces light, it is referred to as an *illuminator* and employs a photodetector as the *detector* (see "chromatography detector" on Wikipedia). The illuminator may emit visible, ultraviolet, or near-infrared light, analysis of which is collectively referred to as *UV / Vis spectroscopy* (see "microspectrophotometry" on Wikipedia). The photodetector absorbs this light and outputs a small electrical current. The electrical current needs amplification and can be presented by using the output voltage to indicate signal strength. A typical voltage output from an industry-standard UV / Vis spectroscopy detector has multiple voltage ranges, with the narrowest range (the highest *sensitivity*) between 0 and 0.01 volts and the widest range between 0 and 5 volts.

Figure 7-1. Experimental setup showing column filtration using the UV / Vis spectrophotometry monitor from Amersham Biosciences. The custom data acquisition and plotting system described in this paper is connected to the UV / Vis monitor output.

Measuring a signal as it is generated is known as *real-time data acquisition*, and visually presenting it is referred to as *plotting*. Both aspects of handling data are needed in many types of experiments, whether in the laboratory or in the field. Additionally, it is typically desirable to store a recording of the signal measurements for either later analysis or archiving, which is referred to as *data logging*. In a biology context, these tasks are performed by a *chart plotter* (see "chart recorder" on Wikipedia), an expensive piece of equipment unavailable to our lab. In an engineering context, these tasks are performed either by a data acquisition board inserted into a computer workstation or by a large-memory digital oscilloscope, also expensive pieces of equipment unavailable to us.

The experiment, shown in Figure 7-1, requires extraction from a solution based on the size of the target, a DNA macromolecule. This is accomplished with *gel filtration chromatography* via a resin-filled *filter column*, which biases the filter output based on particle size as the solution is pumped through the column (this is called *size-exclusion chromatography*). The filter output is routed through an illuminator and detector to generate a detection signal. The signal voltage changes in proportion to particle size. When a critical voltage level occurs, corresponding to the solution containing the desired target macromolecule, the solution is routed using a Y-valve into a designated target container. The noncorresponding solution is routed into a waste container. The success of this method of extraction depends on a low-latency real-time measurement as a liquid flows through a channel. In the experiment shown, the liquid routing was performed manually as the experiment progressed by watching the chart-plotting application window (Figure 7-2), although the data acquisition system could also support controlled automation (see Table 7-1).

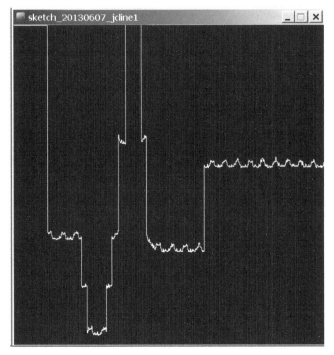

Figure 7-2. A screenshot of this paper's graphical software application during an experiment to test the rails for full-scale operation.

Referring to Figure 7-1, the peristaltic pump (left) transports a solution from a source vial (front) to a destination vial (rear) through the UV illuminator-detector (middle), which is connected to the UV illuminator-detector control unit (right). The UV-detector control unit supplies a signal voltage output at the rear of the enclosure. This voltage output is connected to the USB microcontroller hardware from this paper (hidden from view). The filter column (to the right of the pump), normally in series with the peristaltic pump, is not yet connected.

Engineering Requirements

From the preceding discussion, the following requirements are assumed for plotting the chromatography's illuminator-detector signal. These requirements are comparable to those needed for many biochemistry experiments.

Table 7-1. Engineering requirements

Feature	Requirement discussion
Vertical axis resolution and range	At least 8 bits (256 steps). No more than 12 bits needed (4,096 steps).
	Input must be scaled to 5 volts. Maximum low-sensitivity input signal is 0.1 volts (required gain of 50). Maximum high-sensitivity input signal is 10 millivolts (required gain 500). Minimum voltage is 0 volts.
Horizontal axis resolution and tolerance	Maximum latency: 200 milliseconds; fast enough to control a valve.
	Maximum jitter: 2 milliseconds.
Sampling rate	1 to 10 Hz (sample period of 0.1 seconds).
Data bandwidth and data logging memory size	12 bits at 2 Hz (very low data rate). Any modern flash memory is sufficient; for simplicity of handling data, the data storage/archiving is performed on the computer, so data logging of billions of samples is trivial.
Peripheral communication	Basic USB 1.1 communication as a slave device would allow the peripheral to be recognized as a simple serial data communications device, requiring no special software. Features unique to USB 2.0 and later provide no significant benefits. A drawback to using USB is that the ground plane is shared between the master and the peripheral device, opening slight possibilities of damage to the computer or introducing noise in the signal measurement.
	A wireless technology (Bluetooth or WiFi) could be used, but this introduces additional cost and complexity and should be avoided.
Power	If using USB communication, the power is 5 volts from the USB master. In this case, analog circuits use a single-sided supply. Circuits powered via USB consume less than 500 mA.
	If using wireless, then power sources either from a supply or a battery pack providing at least 6.5 volts.
Digital interfacing to control liquid automation (I/O)	A communication channel for control of digital I/O, and at least one pin of controllable digital output pin (for future use). A digital output pin can control a liquid flow valve, for example.

Feature	Requirement discussion
Computer operating system application compatibility	Compatible with all common, modern OSs (Mac OS X, Microsoft Windows, GNU/Linux, and BSD Unix) and standardized embedded computers such as the Raspberry Pi and MIPS or ARM boards (GNU/Linux with X-Windows).
	For compatibility across these, and to allow modifications by non-computer scientists for simple customizations, there are two modern software platform choices: Processing (see the sidebar; runs on Java) or Python with Qt.
	Application software, and the required underlying software, must be simple to install and simple to run. The software interface is best focused on immediate visualization of the real-time data.
DIY project complexity	The custom hardware can be built by anyone aged eight or older. Off-the-shelf components are used.
Standards and certifications	None. DIY designs need not be certified as medical grade, although a peer-reviewed open hardware and open source design may meet similar levels of quality in practice.

About Processing

Processing is an application programming system originally developed for non-computer scientists (refer to *processing.org*). It is a simplified environment in which to develop applications that display graphical results. Processing is also used as the basis for the Arduino programming system; however, as the Universal Bit Whacker (see the upcoming sidebar) is a better fit to most requirements, using Arduino hardware is not recommended.

Design

This paper describes several technological building blocks for creating a basic data acquisition system. Major components are highlighted in Table 7-2.

Table 7-2. Building blocks of this design

Component	Purpose	Technical reference
Operational amplifier	The op-amp amplifies the signal from the detector. The LM358 op-amp can be powered by the 5 V provided by the USB connection and contains dual amplifiers in a single chip, as shown in Figure 7-4. To simplify the design for those new to the concept of op-amps, positive feedback is used instead of negative feedback.	LM358 datasheet from ST Microelectronics.
Custom analog circuitry	A bias circuit that uses the op-amp buffers and scales the laboratory instrument signal for connection to the microcontroller. The circuit uses the non-inverting configuration of the operational amplifier to simplify understanding of the software.	*The Art of Electronics*, by Paul Horowitz and Winfield Hill (Cambridge University Press), and *Analog Integrated Circuits Applications,* by J. Michael Jacob (Prentice Hall).
Microcontroller board	Used for data acquisition, USB communication, and circuit power. The Microchip PIC18 microcontroller provides USB support and a built-in 12-bit analog-to-digital converter.	The UBW project (*http://bit.ly/bio6-ubw*) by Brian Schmalz, and the Microchip PIC18 datasheet and family guide from Microchip. The UBW board is commercially available with documentation from SparkFun.
Processing programming environment	Software system for running the custom application. *Processing* allows non-computer scientists to write simple applications that include plotting and graphics functionality.	*Processing.org.* (*http://processing.org*)

Component	Purpose	Technical reference
The custom source code in this paper, the Chart Recorder Application	Communicates with the UBW hardware and plots visual data.	This paper, and Processing examples.
Copper bus trace prototyping board ("breadboard" or "protoboard")	Using a pre-drilled FR4 board with copper bus traces helps in construction of electronic circuits. Copper bus boards are very handy for DIY, so buying a few extra is recommended.	Radio Shack part #276-168 or #276-150.

The Universal Bit Whacker (UBW) Microcontroller Kit

The Universal Bit Whacker (*http://www.schmalzhaus.com/UBW*), shown in Figure 7-3, is an open hardware and open software controller developed at the University of California at Santa Barbara. The UBW accepts simple text-based commands to toggle digital output pins, read digital input pins, or read from the analog-to-digital converter on the analog input pins. When connected via USB, the UBW presents itself as a simple serial data communications port. The UBW may also be connected to the Raspberry Pi embedded computer board as a USB peripheral. For more advanced use, the UBW firmware is open source and available for modification, or completely new firmware may be written. The standard UBW firmware is written in C. A free drag-and-drop application is available to update the microcontroller's flash program memory (the application is available for Mac OS X and Microsoft Windows). The Microchip PIC may also be programmed directly with a low-cost chip programmer. The UBW firmware is compiled with an integrated software development environment (IDE), MPLAB/X, which includes a free C compiler for all PIC microcontrollers. MPLAB/X is compatible with Mac OS X, Microsoft Windows, and GNU/Linux; it is available from Microchip, Inc.

Figure 7-3. The USB microcontroller circuit board, a Universal Bit Whacker (UBW).

A Summary of Operational Amplifiers and Circuit Characterization for Biologists

An operational amplifier (tersely, an op-amp), summarized in biology terms, is an interconnection of semiconductors arranged in a complex network with specific biasing along each pathway in order to amplify an input signal. The network includes feedback paths that result in particular performance characteristics and also create stability under normal conditions. This may be viewed as similar to a metabolic network. The output of the network is a linear amplification of the input signal, both given as voltages and related to the power voltage supplied to the op-amp on the power path.

The various internal biasing of each network path has been characterized and fine-tuned by thousands of interns, graduate students, and post-docs over many decades, often by trial and error, and often after very long hours in the laboratory. Semiconductor manufacturers produce ready-made, pre-biased networks as operational-amplifier integrated circuits (nicknamed *chips*) with well-known and standardized part numbers.

The standard LM741 op-amp, for example, made by several manufacturers, has been continually studied and characterized, especially by third-year electrical engineering undergraduates, since the 1970s. Op-amp chips provide standardized external contact pins for connecting bias components, such as resistors, capacitors, or inductors, chosen by a board-level design engineer (or student). These external contact pins connect into the internal feedback network. The characterization is so well-studied that commonly used operational amplifier chips require only a single resistor on one path, a bare wire for another path, and two capacitors on their respective paths for a completely stable network with linear response.

Internally, thousands of circuit connections provide voltage, current, and temperature feedback to maintain a steady bias for the design engineer's desired output. Operational amplifiers have some interesting characteristics as a result of the bias network. The network allows the input signal to be connected as part of either a positive feedback loop or a negative feedback loop. The most stable configuration employs negative feedback, which, as a side effect, inverts the input when generating the output (for a single-ended op-amp, if the input is 2 V, the output will be $V_{max}-2$ V). If the external bias components are not chosen properly (or are purposely chosen improperly, for experimentation), the operational amplifier will demonstrate instability, meaning that the output may hit the maximum or minimum voltage available (commonly referred to as hitting the rails or clipping), or may oscillate rapidly, which creates noise.

Semiconductor networks are currently more predictable than biology's metabolic networks, due to the large volume of work that has been done in characterizing their behavior; however, naturally evolved metabolic networks, when compared to semiconductor networks, are more self-stable (until the environment kills the metabolic network, which may involve some noise as well, from the biological host).

For the purposes of the device in this paper, the critical importance of the operational amplifier is its use as an electron buffer. Under correct bias, the input to the operational amplifier steals very little current from the output of the spectroscopy detector, such that the detector's output is not modified as a result of observation; this is referred to as *impedance matching* or as a circuit connection having *high impedance*. Operational amplifiers with an amplification of unity (gain of one) are frequently used as the first interconnecting stage to buffer the input. For an inverting configuration, the gain is given by $G = -R_F/R_n$. For a noninverting configuration, the gain is given by $G = 1 + R_F/R_n$. The inverting input is denoted on the symbol of an op-amp with a minus sign, and the noninverting input is denoted with a plus sign (refer to Figure 7-4).

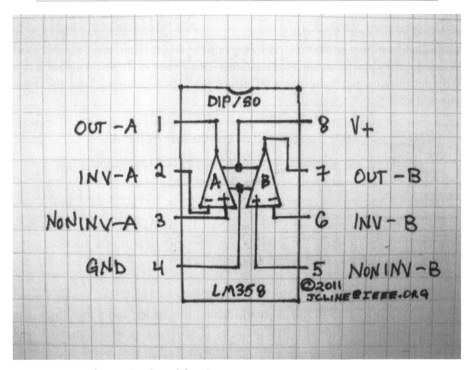

Figure 7-4. LM358 operational amplifier pinout.

The gain of the operational amplifier is matched to the laboratory instrument to result in a full-scale o to 5 volt range at the UBW pin. To adapt the design to other laboratory instruments, measure the voltage output of the instrument and choose new resistor values according to op-amp gain equations (see sidebar) that yield a maximum 5 volt output to the UBW. The feedback resistors should be kept within the range of 50 k Ω to 5 M Ω for proper performance.

The UBW is powered simply by plugging it in to the computer via USB. This should be performed before starting the application software. The application makes a small effort to discover if the UBW is connected to the computer by listing all detected USB devices, sending a Version command to a specific device on the list that is assumed to be the UBW, then reading the reply to detect the response from the UBW. On OS X or GNU/Linux operating systems, the device enumeration is the device driver's name as assigned by the operating system, a simple serial TTY (such as /dev/ttyS0). This device name is held constant regardless of which physical USB port the UBW is connected to; therefore, the application is modified only once and accesses the correct device from then on. On Microsoft operating systems, the device name will be an enumerated COM port, assigned by the operating system depending on which physical port the UBW is connected to (thus confusingly changing enumeration). Therefore, on Microsoft Windows, the application must be modified each time the UBW is connected to a different physical port; alternatively, ensure the UBW is plugged into the same physical USB port for each use.

Source Code for Chart Recorder Application

The source code to accompany this paper is available at the 88 Proof Synth Bio Blog (*http://bit.ly/bio6-88p*).

When first connected, the UBW waits in idle operation. Pressing the UBW's Reset button also restores the UBW to idle operation. From idle operation, the UBW needs setup commands for beginning acquisition. Pins RA0 and RA1 are used on the UBW for analog input in this circuit and are referenced in the commands. First, enable the proper UBW analog pins for analog input (C,255,255,255,1 and CU,1,0). Second, enable the analog sampling timer at a chosen sampling rate, given in milliseconds (T,200,1), after which the UBW will output the analog readings, one pin's conversion per text line. Third, begin reading the UBW analog conversion data in a loop. The application software automatically performs these steps.

The application, as listed, ends acquisition after a fixed number of readings (20,000 samples). In the given experiment, this acquisition length corresponded to a full cycle of filtration. An improvement in the application would be to include user-interface controls such as start, stop, and pause to control acquisition and which additionally control the peristaltic pump, using the UBW digital output pins with the digital controls of the pump. An improved application could also detect sample readings that hit the rails, such as when liquid flow is interrupted, to indicate that user intervention may be required.

Building the System

The circuit is simple enough to construct directly by soldering point-to-point onto a copper prototyping board (see the schematic, Figure 7-5). It is best to spatially separate analog circuits from digital circuits to reduce electrical noise. In general, analog circuits are best soldered onto a copper board rather than inserted into a solderless breadboard to reduce stray capacitance. Leads of components are trimmed to reduce length and reduce stray inductance. A copper prototyping board with bus traces is used to ease construction (see Figure 7-6).

The recommended bus board provides five connections for each pin in a dual in-line package (DIP) integrated circuit (IC) footprint, with two power rails down the center. The power rails are helpful in higher-speed digital circuits, to reduce digital noise, or in analog circuits like this one, as a ground bus. In constructing this board, the two power rails are soldered together at a single point and used as a larger ground. The bypass capacitor in particular, used to reduce power supply noise, should have short leads and is often arranged diagonally across the top of the IC to reach as close to the IC as possible while connecting power to ground.

The UBW, with male headers, is plugged into the board's female headers (see Figure 7-7).

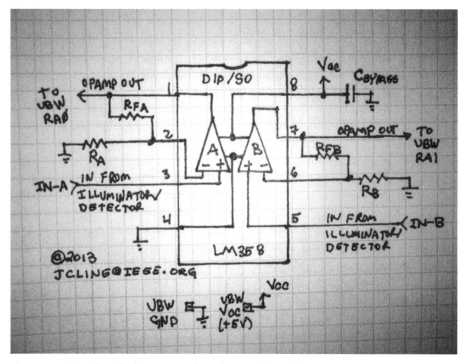

Figure 7-5. Circuit board schematic. Refer to the parts list (Table 7-3).

Table 7-3. Parts list

Quantity	Cost	Component
1	$25	UBW preassembled circuit board with male headers, SparkFun.com part #DEV-00762.
1	$2	USB mini device cable (the standard cable used by many USB peripherals).
1	$0.50	LM358 single-ended dual op-amp, Texas Instruments. Available at Digikey or a local electronics retailer.
1	$3	Copper prototyping board, Radio Shack part #276-168 or #276-150.
1 set	$3	1/8 W, 1% resistors and capacitors: $C_{bypass} = 1$ uF, $R_A = 2$ k Ω, $R_{FA} = 100$k Ω, $R_B = 2$ k Ω, $R_{FB} = 1$ M Ω. 0.1" female headers for the board accept the 0.1" male headers on the UBW.

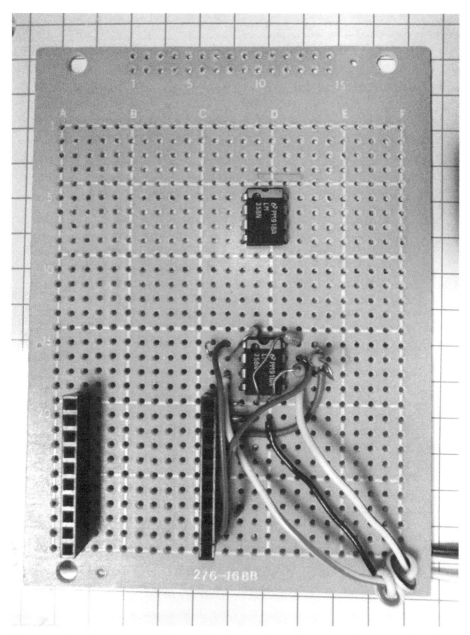

Figure 7-6. Simple custom circuit board of the device built in this paper, with the operational amplifier and bias circuit shown (bottom). The UBW plugs into the female headers at bottom left. The board is constructed with point-to-point soldering.

Figure 7-7. Complete custom circuit board of the device built in this paper, with an operational amplifier and USB microcontroller.

Using or Building Custom USB-Powered Peripherals

NOTE

A powered USB hub is recommended between the laptop and any custom USB microcontroller circuit for protection. The USB peripheral is coupled into the computer power bus, and isolation is preferred. A powered USB hub may provide better isolation (though not always). Also, the USB peripheral may accidentally exceed the maximum current capabilities of the computer's USB port. A powered USB hub may provide better overcurrent power protection in case of accidental shorts or construction errors. At a minimum, burning up a powered USB hub is preferable to burning up a computer's motherboard.

Conclusion

This simple data acquisition system functioned properly during the experiment described and is now a tool in professional laboratory use. It was constructed within the cost and time constraints described. The concepts presented in this paper should allow modifications for functioning in a variety of experiments where data acquisition and chart plotting are needed. A recommended future improvement is to add data-logging features into the application software by recording timestamped data in a comma-separated value file format (*.csv*); also, an autodetection feature in the software that scans all connected serial ports to automatically locate the UBW's port would be helpful to laboratory users.

Jonathan Cline provides his personal genome sequence data for immediate download as a participant in the Personal Genome Project and as proof he is not a zombie. He invites readers to hack his genome, available at http://88proof.com/about. He holds a degree in electrical engineering and specializes in embedded systems software and hardware, network protocols, and cryptography. Jonathan has proposed and demonstrated several biohacking projects. He can be reached via email at jcline@ieee.org. Digital copies of the engineering resources in this paper are available via email.

DIY Paper
Microfluidics

Sim Castle

Microfluidics, the practice of manipulating microliter volumes of liquids, is a hot topic in research at the moment, and for good reason. The ability to deal with tiny quantities of analyte opens up a whole new realm of possibilities: from the study of individual cells to the Wyss Institute's "organ-on-a-chip (*http://bit.ly/bio6-wyss*)" devices, which promise a way of studying the function of human organs in vitro.

Despite the relative simplicity of these devices, they require a range of equipment to run, from pumps and control instruments to microscopy equipment. On top of this, the manufacturing of these devices is fairly complex and requires specialist equipment. All of this puts microfluidics devices out of range for the average DIY biologist.

Paper microfluidics devices offer a more promising solution in this regard: although the same degree of control of fluid flow cannot be achieved (as flow is simply governed by the capillary action in the paper), these devices can be manufactured with no specialist equipment and with minimal cost. Despite their simplicity, paper microfluidics hold great potential for various applications, perhaps most notably for field diagnostics. A paper-based Ebola sensor has been developed at the Wyss Institute and works by embedding an RNA-based genetic circuit directly into the paper itself.[1] Applications of paper microfluidics reach beyond just sensors: they have been used to separate plasma in blood samples[2] and as a substrate for growing three-dimensional cell cultures,[2] to name just a couple of the innovative uses of this simple technology.

All paper-microfluidics devices basically work through the same principle: most of the paper is hydrophobic in some way, except for a specified area that is left with the regular hydrophilic properties of filter paper. This area serves as the microfluidics channels. Fluid is then drawn up the channels through capillary action.

There are several methods that can be used to fabricate paper microfluidics devices, most of which are within reach for the average DIYer. The most commonly used method, and the one I have had most success with, is photolithography.[4] The basic working principle is to saturate a piece of filter paper with photoresist, a special type of UV light–activated epoxy, mask the areas where you want to create your fluid channels, and expose the paper to UV light. Once developed, you have a combination of hydrophobic areas (where the epoxy has set due to exposure) and hydrophilic areas (where the epoxy has been washed away) that make up the device.

Materials

Filter paper
> I've tried various types of paper and found thin paper with fine fibers to work best. Thicker grades are not suitable because the epoxy does not fully soak through.

UV-activated epoxy
> SU-8 is the lab-grade compound most often used and can be ordered from Sigma-Aldrich. However, UV epoxy is also used to make craft jewelry and fishing tackle and can be readily bought in smaller quantities for such purposes, for only a few dollars.

Small brush or Q-tip
> For applying the epoxy.

UV light
> A source of UV light is needed to activate the resin. This needs to be close to 365 nm. I used a fancy UV light box that I borrowed from my university, but a UV flashlight should do the job, provided it's below about 390 nm. If you are blessed with great weather, you can even leave it in the sun for 5 to 10 minutes.

Ethanol
> Rubbing alcohol or similar.

Acetone
> Nail polish remover will do the job.

Hot plate
> I just used my ceramic stovetop for this.

Transparency sheets and inkjet printer or permanent marker
The transparencies used need to match the type of printer you use. Laser printers and inkjets require different types of sheets.

Microscope slides and clips
These are to clamp together the mask and the paper for UV exposure.

Method

Figure 8-1. The masks are the patterns with which you create your desired channels. These can easily be created by printing black ink onto your transparency sheets. If dithering is still visible and the printed area is still not entirely transparent, run it through the printer another time (although this will reduce the precision of the mask). Any graphic design or CAD software can be used to create the masks; I used Adobe Illustrator. Alternatively, for a lo-fi pattern, a black permanent marker will do the job, as demonstrated here. As with using a printer, the masked area should be as opaque as possible, so apply several layers if necessary.

Figure 8-2. Apply epoxy to the filter paper and brush it on until it is fully saturated. Warm the epoxy slightly on a radiator; if it is too viscous, wipe off excess. Photoresist is not toxic; however, as with any solvents, care should be taken to use it in a well-ventilated space.

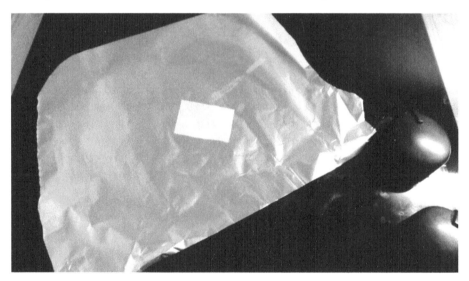

Figure 8-3. Allow to dry, either on a hot plate (about 80 degrees Celsius; I used the lowest setting on my stove) or under ambient conditions.

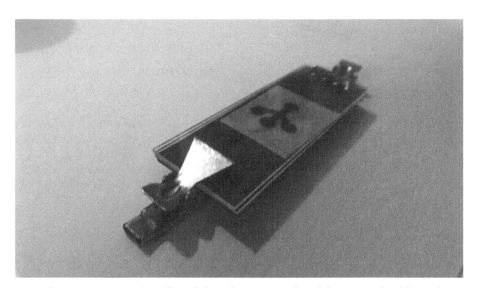

Figure 8-4. Cover paper with mask, and clamp between two glass slides. It's worth adding a sheet of black paper behind the filter paper to reduce UV light leakage.

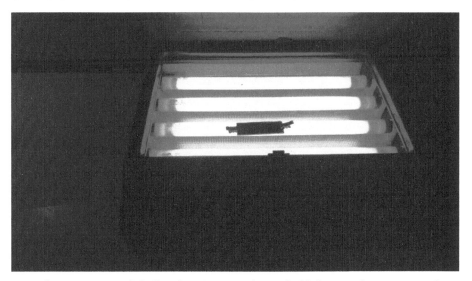

Figure 8-5. Expose to UV light for a few minutes, with your flashlight or UV lamp. You may have to experiment with exposure time, depending on your light source. I used 5 minutes to be sure. (If you're using sunlight, leave it somewhere bright for about 10 minutes; the following heating step is not needed in this case.)

Figure 8-6. Remove paper from slides, and place paper on a hot plate (about 80 degrees Celsius; I used the lowest setting on my electric stovetop) for around 5 minutes. A color change of the unmasked area should occur, and your pattern should become visible. Depending on the epoxy you use, this color change may only be very slight. This final baking step finishes off the polymerization process of the resin.

Figure 8-7. Soak in acetone (I used nail polish remover, which is only available in bright pink for some reason) for a minute or so to wash off the unset epoxy. You should see the color change become more pronounced, and your pattern will take form. Rinse off the acetone with ethanol. Dry under ambient conditions or on the hotplate for 5 minutes. Again, make sure to have a well-ventilated space when using acetone, particularly when drying on the hot plate.

These devices can be used for simple diagnostic tasks, as an alternative to microwell plates, or even for more advanced applications, such as the Ebola test described earlier. 3D devices have also been made by stacking several sheets of paper, opening up even more possibilities.[5] Once paper has been soaked in photoresist and dried, it can be stored indefinitely (in a cool, dark place), so it's easy to make these devices quickly and easily in large batches, if needed.

If you're thinking of using paper microfluidics in your project, I'd love to hear about it! I can be reached on Twitter *@simcastle*.

Figure 8-8. The device is now ready to use. The two on the left have loading die applied and clearly show the channels. The difference in color depending on the resin you use is also seen here: lab-grade SU-8 creates results in a darker color, whereas the commercial epoxy produces very little color, making the channels hard to see.

References

1. Keith Pardee et al., "Paper-based Synthetic Gene Networks," *Cell* 159 (2014): 940–954.

2. Xiaoxi Yang et al., "Integrated separation of blood plasma from whole blood for microfluidic paper-based analytical devices," *Lab Chip* 12 (2012): 274-280.

3. Ratmir Derda et al., "Paper-supported 3D cell culture for tissue-based bioassays," *PNAS* 106 (2009): 18457-18462.

4. George Whitesides et al., "Diagnostics for the Developing World: Microfluidic Paper-Based Analytical Devices," *Analytical Chemistry* 82, no. 1 (2010): 3-10.

5. George Whitesides et al., "Three-dimensional microfluidic devices fabricated in layered paper and tape," *PNAS* 105 (2008): 19606-19611.

Sim Castle is an industrial design engineering graduate student at TU Delft who is exploring the applications of biology in future products. Find him on Twitter @simcastle.

#ScienceHack SynBio Hack-a-Thon

Connor Dickie

Last March at SXSW Interactive, we announced #ScienceHack, a collaborative, open-science effort to make real medicine using synthetic biology. #ScienceHack was born out of the idea that with today's technology, it should be cheap, quick, and easy to create a life-science hack-a-thon that would bring people together for education, innovation, and the development of new materials, medicines, food, and fuels.

Hack-a-thons have been a global phenomenon, exploding in number over the past 15 years. This is possible because computers and free, sufficiently user-friendly development platforms like Linux and Arduino are readily available. Hack-a-thons attract innovative and risk-tolerant people eager to build something cool and maybe stay up all night doing it. As a result, hack-a-thons have become a breeding ground for new start-ups, IP, and art.

#ScienceHack is a series of individual life-science hack-a-thons that are connected by open-science methodologies, web-based tools, and an integrated, common wetware kit. Each #ScienceHack includes six hours of DNA design and build, split equally across two days, and at least one follow-up session to view and record results. The goal is simple: design and build the most efficient and robust violacein-producing *E. coli* colony the universe has ever seen, learn a lot, and have fun!

What Is Violacein?

Violacein is a naturally occurring dark violet material that has shown promising antibiotic and anti-cancer properties. It's produced in nature by *Chromobacterium violaceum*, which can be found living in Amazonian soil. Currently, research-grade violacein is sourced solely from Sigma-Aldrich, and comes with a hefty price tag of $3,500 per milligram (that's $13.3 million per ounce). This exorbitant price keeps this compelling compound out of researchers' hands, which may have a chilling effect on the development of therapies based on violacein.

The genetics of *C. violaceum* is relatively well understood. It was first sequenced in 2003, and in 2009 it was first used in the iGEM community by the University of Cambridge as part of the *E. chromi* project. Thanks to the open-science methodologies employed by the iGEM community, we were able to quickly and inexpensively create and ship an enthusiast-focused wetware product on top of their research.

Violacein Factory

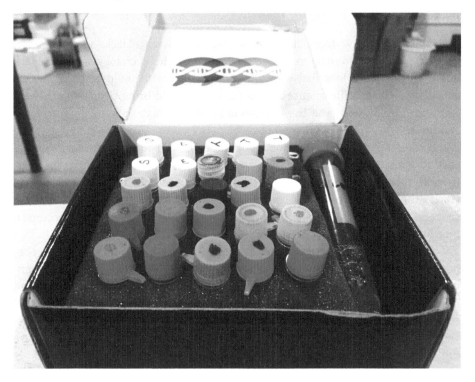

Many people's first experience with a wetware kit is through iGEM. Each year, as part of registration, teams are sent a large collection of DNA parts that have been pulled from the broader BioBricks DNA code registry. In laboratories, iGEM participants assemble the kit parts in unique ways and even incorporate custom parts to invent new organisms that do cool things, all with the hopes of winning the Golden Brick award at the Giant Jamboree. Thanks to participant submissions, the registry has grown to more than 10,000 BioBricks over the past decade, making it an incredible resource for anyone who is interested in synthetic biology development.

My first experience with a wetware kit was through different channels. In the summer of 2012, the FBI held a two-day DIYbio/FBI meet-up (cocktails with the Department of Defense as well) in Walnut Creek, California. One of the breakout sessions had us trek to BioCurious in Sunnyvale, California, where we got our nitrile-clad hands on a prototype wetware kit that the FBI called disruptively simple, fast, and effective.

That amazing wetware kit turned out to be the Rainbow Factory from Genomikon, itself stemming from the 2009 iGEM competition. What's really exciting about the Genomikon assembly standard is that all parts come ready to assemble right out of the box, and assembly is accomplished with magnetic-bead technology. This assembly process allows enthusiasts to create plasmids exceeding 10,000 base pairs in about two hours, using a simple process that requires the use of minimal lab equipment, meaning anyone can build custom DNA parts practically anywhere in just a few hours.

After assembling my very first plasmid that day, it was clear to me that Genomikon, or something similar, would bring hacking biology at the molecular level to a much broader audience of people who are excited about finding solutions with synthetic biology. With the Genokimon assembly standard it is now possible to design and build a novel organism inside of 48 hours, a key component in enabling the possibility of life-science hack-a-thons.

We created the Violacein Factory by converting a collection of violacein-encoding BioBricks into Genomikon standard parts and packaged them as a kit that we made available to the general public on the Synbiota marketplace. You could say that we "ported" one standard to another, giving parts from the old library upgraded features. The final product sells for $450 and has enough DNA to create about 20 plasmids.

In addition to all the physical DNA parts and supporting materials, Violacein Factory also comes with a special project on Synbiota (a free, collaborative virtual laboratory platform that allows groups to easily set up and manage life-science projects on the Web) that includes all the DNA parts as objects ready to manipulate in GENtle, a web-based open source DNA design tool. Using GENtle, people can build computer models of their plasmids using a simple pick-and-place interface.

DNA Design (In Silico)

Duration: one hour

While the Violacein Factory kit is presented as an enthusiast-level product, it is robust enough to be used for real research. It's possible to simply follow the step-by-step instructions to build a reference plasmid and get an organism that produces violacein. More adventurous participants also have the ability to assemble a custom plasmid by modifying the order of the parts and/or adding multiples of parts. This modularity-enabled flexibility, coupled with the quick design and build time, has turned out to be a powerful prototyping combination.

Designing a custom plasmid model is simple. In a #ScienceHack project on Synbiota, a participant navigates to the Sequences tab and opens all 16 #Science-Hack DNA parts in GENtle. Using the designer, a model plasmid is created piece by piece by choosing one part after the other. This modeling process results in a set of instructions that can be used to build a physical plasmid using Genomikon parts by hand (or, in the near future, automatically built using liquid-handling robots and microfluidics).

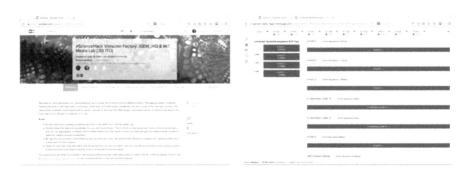

Figure 9-1. #ScienceHack project on Synbiota and designing a plasmid in GENtle

Each participant's plasmid designs are saved in the project's Lab Book. There is also room here for adding additional content, including hypotheses, results, data, and files, making everything well organized, searchable, and immediately available to other #ScienceHackers around the globe.

Assembly Protocol (In Vitro)

Duration: two hours

ASSEMBLE DNA

1. Add 5 uL of the DNA part to washed beads.

2. Add 5 uL of ligase.

3. Gently stir with your pipette or gently flick.

4. Leave at room temperature for 10 minutes. You can gently flick the tube after 5 minutes to mix. Do not flick so hard that droplets fly onto the side of the tube.

5. Wash two times.

6. Repeat with the rest of your DNA sequences up until the last part, which must be the Z--Chloramphenicol_acetyltransferase--CAP part, and make sure you wash twice after your 10-minute incubation.

7. You have now added all your DNA parts, including the last CAP part that has chloramphenicol resistance. You now need to remove your DNA from the magnetic bead so it can be put into a bacteria. This next step is called *elution*.

   After you have added the Z--Chloramphenicol_acetyltransferase--CAP and washed twice, you must then separate your DNA from the bead. Once free in solution, the anchor then binds to the CAP and circularizes your DNA into a plasmid. This finalizes your assembled DNA construct.

ELUTION

1. Add 20 µL of elution solution to bead pellet.

2. Pipette up and down several times to re-suspend the beads and wait 30 seconds.

3. Hold magnet against tube to pull the beads to the side and wait until the solution fully clears (~30s).

4. Keep all liquid, including any droplets on side of tube. This liquid contains your DNA construct! Transfer it into a new PCR tube with your name and a "(–)" on it!

5. Your DNA can be used to transform bacteria and run in an agarose gel.

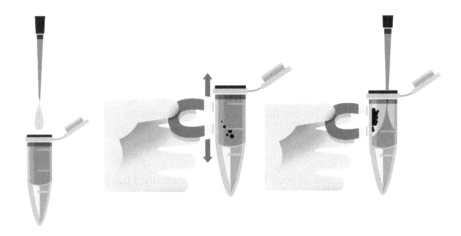

Figure 9-2. Magnetic bead DNA assembly

Transformation Protocol (In Vivo)

Duration: three hours

MATERIALS

1. Your assembled DNA.

2. Transformation buffer.

3. *E. coli* grown the previous night.

4. P20 & P200 pipette and tips.

5. Epitubes.

6. Ice and ice bucket.

7. 42-degree-Celcius water bath or container with 42-degree-Celcius water (you'll need a thermometer in this case).

8. Control DNA if you're running a control.

9. LB media.

10. Chloramphenicol agar plates.

11. Spreader.

12. 37-degree-Celcius incubator.

PROCEDURE

1. Label one epi tube (+) and another (–). Label both with your group name and ensure they are on ice.

2. Label your (+) and (–) chloramphenicol agar plates with your team name.

3. Add 100 uL of transformation buffer (T) to each tube and place on ice.

4. Scrape about 6 small TG1 colonies with a pipette tip and gently suspend the bacteria in the transformation buffer by swirling and gently pipetting up and down. Do this for both tubes.

5. Gently and slowly add 10 uL of your assembled DNA (–) and 2 uL of control DNA (+) to the appropriate tube of competent cells while slowly stirring.

6. Let the samples stand on ice for 30 minutes.

7. Place the samples in a 42-degree-Celcius water bath for 90 seconds.

8. Immediately place back on ice for five minutes.

9. Add 300 uL LB media to each epi tube.

10. Incubate in a 37-degree-Celcius incubator for 30 minutes, then get your labelled agar plates.

11. Pipette 100 uL of the experimental assembled cells onto one (–) plate and 100 uL of the control cells onto the (+) plate.

12. Spread out the bacteria on your plates using a disposable plastic spreader or glass beads.

13. Incubate the plates facing up for 20 minutes (to allow for evaporation of liquid media) and subsequently flip the plates upside down and incubate at 37 degrees Celcius.

Figure 9-3. Streaking transformed E. coli on growth media

Current Status

Figure 9-4. #ScienceHacking in Joi Ito's kitchen in Cambridge, Massachusetts.

#ScienceHack is an ongoing effort, but since its launch last spring there have been 11 #ScienceHacks at many different types of venues, including hotel bathrooms, Joi Ito's kitchen, Genspace (with help from an OpenTrons robot), and a fine arts class at UPenn with students who had never used a pipette before.

As of writing this article, the #ScienceHack team with the most advanced results is from Ryerson University, in downtown Toronto. Team leader Alejandro Saettone, a master's student in life science, has not only produced a number of very strong violacein-producing colonies, but also developed an inexpensive violacein extraction technique that will enable the community to create high-quality violacein at home.

Alejandro did not stop there. He is testing the team's violacein on both mouse and human cancer cells, as well as comparing their violacein against the research-grade product from Sigma.

For those who are curious, Alejandro has embraced the open science model and shares all of the team's results on Synbiota online (*http://bit.ly/bio6-synbiota*).

#ScienceHack by the Numbers

- Active for 8 months
- 11 #ScienceHack events
- 140 participants, more than half without prior experience
- 10 cities
- 4 countries
- 129 DNA sequences designed and assembled
- 162 Lab Book entries created

Explore all of the #ScienceHack projects and data on the Synbiota website (*http://bit.ly/bio6-scihack*).

What's Next

A lot has happened since we announced #ScienceHack, but much remains to be done now that we have a promising violacein producing candidate. Next on the list is quantification, which will involve not only sequencing the DNA of S6 (to confirm that it is indeed what was designed in GENtle) and the other top violacein-producing candidates, but also quantifying exactly how much violacein is produced by these colonies. This will close the typical prototyping loop of design, build, grow, and test.

Connor Dickie (@connor) cofounded Synbiota to accelerate biotechnology R&D. Synbiota is a rapid prototyping software/wetware platform that puts the power of life into the hands of scientists and enthusiasts around the globe. Connor is an alumnus of the MIT Media Lab, where he created context-sensitive and attention-aware computers. Connor is also an alumnus of Mozilla's WebFWD program and winner of the 2014 SXSW Interactive Accelerator.

Community Announcements

SynBioBeta Conference 2015

 synbiobeta

SynBioBeta London will be back at Imperial College for its third year in a row on April 22nd and 23rd. The synthetic biology community will meet to hear from the companies releasing new technology, commercializing cutting edge research and getting product to market. Join the entrepreneurs and thought leaders who are infusing new biological technologies into the growing bioeconomy. The event is being hosted by SynbiCITE, the Innovation and Knowledge Centre at Imperial College London that is dedicated to promoting the adoption and use of synthetic biology by industry. More information can be found at *www.synbiobeta.com*.

BioFoundry

You already know BioCurious in San Francisco, Genspace in New York and La Paillasse in Paris; now there's BioFoundry in Sydney! BioFoundry is Australia's first community lab for citizen scientists. With 3D printed parts and cast-offs from university and industry, BioFoundry is the home of BioHack Sydney and is due to host Australia's first high school IGEM team this year. Its mission is to democratize science by breaking the cost barrier to research and education. Come and say hello if you are in town!

FundScience

FundScience began as Australia's first nonprofit crowdfunding platform for science research, training, and innovation. FundScience successfully crowdfunded Australia's first BIOMOD team, echiDNA, who won the BIOMOD Grand Prize last November in their inaugural run. FundScience has also turned out to be about more than just crowdfunding: with the implementation of features that allow researchers to share the progress of their projects, FundScience is due to relaunch as an open science platform soon! Learn more at *fundscience.org.au*.

Lightning Source UK Ltd
Milton Keynes UK
UKOW07f1504230115

245014UK00005B/16/P